Tucholsky Wagner Zola Scott Sydow Freud Schlegel
Turgenev Wallace Fonatne
Twain Walther von der Vogelweide Fouqué Friedrich II. von Preußen
Weber Freiligrath
Kant Ernst Frey
Fechner Fichte Weiße Rose von Fallersleben Richthofen Frommel
Hölderlin
Engels Fielding Eichendorff Tacitus Dumas
Fehrs Faber Flaubert
Eliasberg Ebner Eschenbach
Maximilian I. von Habsburg Fock Eliot Zweig
Feuerbach Ewald Vergil
Goethe London
Elisabeth von Österreich
Mendelssohn Balzac Shakespeare Dostojewski Ganghofer
Lichtenberg Rathenau Doyle Gjellerup
Trackl Stevenson Tolstoi Hambruch
Mommsen Thoma Lenz Hanrieder Droste-Hülshoff
von Arnim Hägele Hauff Humboldt
Dach Verne
Karrillon Reuter Rousseau Hagen Hauptmann Gautier
Garschin
Damaschke Defoe Hebbel Baudelaire
Descartes
Hegel Kussmaul Herder
Wolfram von Eschenbach Darwin Dickens Schopenhauer Rilke George
Bronner Melville Grimm Jerome Bebel
Campe Horváth Aristoteles Voltaire Federer Proust
Bismarck Vigny Barlach Herodot
Gengenbach Heine
Storm Casanova Tersteegen Grillparzer Georgy
Chamberlain Lessing Langbein Gilm
Brentano Lafontaine Gryphius
Claudius Schiller Kralik Iffland Sokrates
Strachwitz Schilling
Katharina II. von Rußland Bellamy
Gerstäcker Raabe Gibbon Tschechow
Löns Hesse Hoffmann Gogol Wilde Gleim Vulpius
Luther Heym Hofmannsthal Klee Hölty Morgenstern Goedicke
Roth Heyse Klopstock Kleist
Luxemburg Puschkin Homer Mörike Musil
Machiavelli La Roche Horaz
Kierkegaard Kraft Kraus
Navarra Aurel Musset Moltke
Nestroy Marie de France Lamprecht Kind Kirchhoff Hugo
Laotse Ipsen Liebknecht
Nietzsche Nansen
Marx Lassalle Gorki Klett Ringelnatz
von Ossietzky May Leibniz
vom Stein Lawrence
Irving
Petalozzi Knigge
Platon Pückler Michelangelo Kock Kafka
Sachs Poe Liebermann
de Sade Praetorius Mistral Zetkin Korolenko

The publishing house tradition has created the series **TRADITION CLASSICS**. It contains classical literature works from over two thousand years. Most of these titles have been out of print and off the bookstore shelves for decades.

The book series is intended to preserve the cultural legacy and to promote the timeless works of classical literature. As a reader of a **TRADITION CLASSICS** book, the reader supports the mission to save many of the amazing works of world literature from oblivion.

The symbol of **TRADITION CLASSICS** is Johannes Gutenberg (1400 – 1468), the inventor of movable type printing.

With the series, tradition intends to make thousands of international literature classics available in printed format again – worldwide.

All books are available at book retailers worldwide in paperback and in hardcover. For more information please visit: www.tredition.com

tradition was established in 2006 by Sandra Latusseck and Soenke Schulz. Based in Hamburg, Germany, tradition offers publishing solutions to authors and publishing houses, combined with worldwide distribution of printed and digital book content. tradition is uniquely positioned to enable authors and publishing houses to create books on their own terms and without conventional manufacturing risks.

For more information please visit: www.tredition.com

The Art of Confectionary Shewing the Various Methods of Preserving All Sorts of Fruits

Edward Lambert

Imprint

This book is part of the TREDITION CLASSICS series.

Author: Edward Lambert
Cover design: toepferschumann, Berlin (Germany)

Publisher: tredition GmbH, Hamburg (Germany)
ISBN: 978-3-8495-2459-3

www.tredition.com
www.tredition.de

Copyright:
The content of this book is sourced from the public domain.

The intention of the TREDITION CLASSICS series is to make world literature in the public domain available in printed format. Literary enthusiasts and organizations worldwide have scanned and digitally edited the original texts. tredition has subsequently formatted and redesigned the content into a modern reading layout. Therefore, we cannot guarantee the exact reproduction of the original format of a particular historic edition. Please also note that no modifications have been made to the spelling, therefore it may differ from the orthography used today.

THE

ART

OF

CONFECTIONARY.

Of the Manner of clarifying Sugar, and the different Ways of boiling it.

Since the Ground-work of the Confectioner's Art depends on the Knowledge of clarifying and boiling Sugars, I shall here distinctly set them down, that the several Terms hereafter mentioned may the more easily be understood; which, when thoroughly comprehended, will prevent the unnecessary [Pg 4] Repetitions of them, which would encumber the Work and confound the Practitioner, were they to be explained in every Article, as the Variety of the Matter should require: I shall therefore, through the whole Treatise, stick to these Denominations of the several Degrees of boiling Sugar, *viz.* Clarifying, Smooth, Blown, Feather'd, Cracked, and Carmel.

To Clarify Sugar.

Break into your preserving Pan the White of one Egg, put in four Quarts of Water, beat it up to a Froth with a Whisk, then put in twelve Pounds of Sugar, mixed together, and set it over the Fire; when it boils up, put in a little cold Water, which will cause it to sink; let it rise again, then put in a little more Water; so do for four or five times, till the Scum appears thick on the Top; then remove it from the Fire and let it settle; then take off the Scum, and pass it through your straining Bag.

Note, If the Sugar doth not appear very fine, you must boil it again before you strain it; otherwise in boiling it to an Height, it will rise over the Pan, and give the Artist a great deal of Trouble.

[Pg 5]

The boiling Sugar to the Degree called Smooth.

When your Sugar is thus clarified, put what Quantity you shall have Occasion for over the Fire, to boil smooth, the which you will prove by dipping your Scummer into the Sugar; and then touching it with your Fore-finger and Thumb, in opening them a little you will see a small Thread drawn betwixt, which immediately breaks, and remains in a Drop on your Thumb; thus it is a little smooth; then boiling it more, it will draw into a larger String; then it is become very smooth.

The Blown Sugar.

Boil your Sugar yet longer than the former, and try it thus, *viz.* Dip in your Scummer, and take it out, shaking off what Sugar you can into the Pan, and then blow with your Mouth strongly through the Holes, and if certain Bubbles or Bladders blow through, it is boiled to the Degree called Blown. [Pg 6]

The Feathered Sugar,

Is a higher Degree of boiling Sugar, which is to be proved by dipping the Scummer when it hath boiled somewhat longer; shake it first over the Pan, then giving it a sudden Flurt behind you; if it be enough, the Sugar will fly off like Feathers.

The Crackled Boiling,

Is proved by letting it boil somewhat longer; and then dipping a Stick into the Sugar, which immediately remove into a Pot of cold Water standing by you for that Purpose, drawing off the Sugar that cleaves to the Stick, and if it becomes hard, and will snap in the Water, it is enough; if not, you must boil it till it comes to that Degree.

Note, Your Water must be always very cold, or it will deceive you.

The Carmel Sugar,

Is known by boiling yet longer, and is proved by dipping a Stick, as aforesaid, first in the Sugar, and then in the Water: But this you must observe, when it comes [Pg 7] to the Carmel Height, it will snap like Glass the Moment it touches the cold Water, which is the highest and last Degree of boiling Sugar.

Note, There is this to be observed, that your Fire be not very fierce when you boil this, lest flaming up the Sides of your Pan, it should occasion the Sugar to burn, and so discolour it.

To preserve Seville-Oranges Liquid.

Take the best Seville-Oranges, and pare them very neatly, put them into Salt and Water for about two Hours; then boil them very tender till a Pin will easily go into them; then drain them well from the Water, and put them into your preserving Pan, putting as much clarified Sugar to them as will cover them, laying some Trencher or Plate on them to keep them down; then set them over a Fire, and by Degrees heat them till they boil; then let them have a quick boil till the Sugar comes all over them in a Froth; then set them by till next Day, when you must drain the Syrup from them, and boil it till it becomes very smooth, adding some more clarified Sugar; put it upon the Oranges, and give them a Boil, then set them by till next [Pg 8] Day, when you must do as the Day before. The fourth Day drain them and strain your Syrup through a Bag, and boil it till it becomes very smooth; then take some other clarified Sugar, boil it till it blows very strong, and take some Jelly of Pippins drawn from the Pippins, as I shall immediately express, with the Juice of some other Oranges: As for Example, if you have six Oranges, after they are preserved as above directed, take two Pounds of clarified Sugar, boil it to blow very strong; then one Pint and half of Pippin Jelly, and the Juice of four or five Oranges, boil all together; then put in the Syrup that has been strained and boiled to be very smooth, and give all a Boil; then put your Oranges into your Pots or Glasses, and fill them up with the above made Jelly; when cold cover them, and set them by for Use.

Note, You must be sure in all your Boilings to clear away the Scum, otherwise you will endanger their Working: And if you find they will swim above your Jelly, you must bind them down with a Sprig of a clean Whisk.

[Pg 9]

To draw a Jelly from Pippins.

Take the fairest and firmest Pippins, pour them into fair Water, as much as will cover them; set them over a quick Fire, and boil them to Mash; then put them on a Sieve over an earthen Pan, and press out all the Jelly, which Jelly strain through a Bag, and use as directed in the Oranges before mentioned, and such others as shall be hereafter described.

To make Orange Marmalade.

Take six Oranges, grate two of the Rinds of them upon a Grater, then cut them all, and pick out the Flesh from the Skins and Seeds; put to it the grated Rind, and about half a Pint of Pippin Jelly; take the same Weight of Sugar as you have of this Meat so mingled; boil your Sugar till it blows very strong; then put in the Meat, and boil all very quick till it becomes a Jelly, which you will find by dipping the Scummer, and holding it up to drain; if it be a Jelly, it will break from the Scummer in Flakes; if not, it will run off in little Streams: When it is a good Jelly, put it into your Glasses or Pots.

[Pg 10] *Note*, If you find this Composition too sweet, you may in the boiling add more Juice of Oranges; the different Quickness they have, makes it difficult to prescribe.

To preserve Oranges with a Marmalade in them.

Pare your Oranges as before, make a round Hole in the Bottom, where the Stalk grew, the Bigness of a Shilling; take out the Meat, and put them into Salt and Water for two or three Hours; then boil them very tender, then put them into clarified Sugar, give them a boil the next Day, drain the Syrup and boil it till it becomes smooth; put in your Oranges and give them a good boil. When a little cool, drain them and fill them with a Marmalade made as before directed, putting in the round Piece you cut out; with the Syrup, some other Sugar, and Pippin Juice, make a Jelly, and fill up your Pots or Glasses.

To make a Compote of Oranges.

Cut the Rind off your Oranges into Ribs, leaving part of the Rind on; cut them into eight Quarters, throw them into boiling Water; when a Pin will easily go through the Rind, drain and put them [Pg 11] into boiling Water, when a Pin will easily go through the Rind, drain and put them into as much Sugar boiled, till it becomes smooth, as will cover them, give all a Boil together, adding some Juice of Oranges to what Sharpness you please; you may put a little Pippin Jelly into the Boiling, if you please; when cold serve them to Table on Plates.

To make Orange-Rings and Faggots.

Pare your Oranges as thin, and as narrow as you can, put the Parings into Water, whilst you prepare the Rings, which are done by cutting the Oranges so pared into as many Rings as you please; then cut out the Meat from the Inside; then put the Rings and Faggots into boiling Water; boil them till tender; then put them into clarified Sugar, as much as will cover them; set them by till next Day; then boil all together, and set them by till the Day after; then drain the Syrup, and boil it till very smooth; then return your Oranges into it, and give all a Boil; the next Day boil the Syrup till it rises up to almost the Top of your Pan; then return the Oranges into it, and give it a Boil; then put them by in some Pot to be candied, as [Pg 12] hereafter mentioned, whenever you shall have Occasion.

To candy Orange, Lemon, and Citron.

Drain what Quantity you will candy clean from the Syrup, and wash it in luke-warm Water, and lay it on a Sieve to drain; then take as much clarified Sugar as you think will cover what you will candy, boil it till it blows very strong, then put in your Rings, and boil them till it blows again; then take it from the Fire, and let it cool a little; then with the back of a Spoon rub the Sugar against the Inside of your Pan till you see the Sugar becomes white; then with a Fork take out the Rings one by one, lay them on a Wire-grate to drain, then put in your Faggots, and boil them as before directed; then rub the Sugar, and take them up in Bunches, having some-body to cut them with a Pair of Scizers to what Bigness you please, laying them on your Wire to drain.

Note, Thus may you candy all Sorts of Oranges and Lemon-Peals or Chips.

Lemon Rings and Faggots are done the same Way, with this Distinction only, that the Lemons ought to be pared [Pg 13] twice over, that the Ring may be the whiter; so will you have two Sorts of Faggots: But you must be sure to keep the outward Rind from the other, else it will discolour them.

To make Orange-Cakes.

Take six Sevil-Oranges, grate the Rinds of two of them, and then cut off the Rinds of all six to the very Juice; boil them in Water till very tender; then squeeze out all the Water you can, and beat them to a Paste in a Marble-Morter; then rub it through a Sieve of Hair; what will not easily rub through must be beat again till all is got through; then cut to Pieces the Insides of the Oranges, and rub as much of that through as you possibly can; then boil about six or eight Pippins in as much Water as will almost cover them, and boil them to a Paste, and rub it through a Sieve to the rest; then put all into a Pan together, and give a thorough Heat, till it is well mingled; then to every Pound of this Paste take one Pound and a Quarter of Loaf-sugar; clarify the Sugar, and boil it to the Crick; then put in your Paste and the grated Peal, and stir it all together over a slow Fire till it is well mixed, and the Sugar all melted; then with a Spoon fill [Pg 14] your round Tin-Moulds as fast as you can; when cold, draw off your Moulds, and set them in a warm Stove to dry; when dry on the Tops, turn them on Sieves to dry on the other Side; and when quite dry, box them up.

Lemon-Cakes.

Take six thick-rinded Lemons, grate two of them, then pare off all the yellow Peal, and strip the White to the Juice, which White boil till tender, and make a Paste exactly as above.

To preserve White-Citrons.

Cut your White-Citrons into what sized Pieces you please; put them into Water and Salt for four or five Hours; then wash them in fair Water, and boil them till tender; then drain them, and put them into as much clarified Sugar as will cover them, and set them by till next Day; then drain the Syrup, and boil it a little smooth; when cool, put it on your Citrons; the next Day boil your Syrup quite smooth, and pour on your Citrons; the Day after boil all together and put into a Pot to be candied, or put in Jelly, or compose as you please.

[Pg 15] *Note*, You must look over these Fruits so kept in Syrup; and if you perceive any Froth on them you must give them a Boil; and if by Chance they should become very frothy and sour, you must first boil the Syrup, and then all together.

To preserve Golden-Pippins in Jelly.

Pare your Pippins from all Spots, and with a narrow-pointed Knife make a Hole quite through them, then boil them in fair Water about a Quarter of an Hour; then drain them, and take as much Sugar as will cover them; boil it till it blows very strong, then put in your Pippins, and give them a good Boil; let them cool a little, then give them another Boil; then if you have, for example a Dozen of Pippins, take a Pound of Sugar, and boil it till it blows very strong; then put in Half a Pint of Pippin Jelly and the Juice of three or four Lemons; boil all together, and put to the Golden-Pippins; give them all a Boil, scum them, and put them into the Glasses or Pots.

[Pg 16]

To dry Golden-Pippins.

Pare your Pippins, and make a Hole in them, as above, then weigh them, and boil them till tender; then take them out of the Water, and to every Pound of Pippins take a Pound and a Half of Loaf-Sugar, and boil it till it blows very strong; then put in the Fruit, and boil it very quick, till the Sugar flies all over the Pan; then let them settle, and cool them, scum them, and set them by till the next Day, then drain them, and lay them out to dry, dusting them with fine Sugar before you put them into the Stove; the next Day turn them and dust them again, when dry, pack them up.

Note, You must dry them in Slices or Quarters, after the same Manner.

To make Orange Clear-Cakes.

Take the best Pippins, pare them into as much Water as will cover them; boil them to a Mash; then press out the Jelly upon a Sieve, and strain it through a Bag, adding Juice of Oranges to give it an agreeable Taste: To every Pound of Jelly take one Pound and a Quarter of Loaf [Pg 17] Sugar, boil it till it cracks, then put in the Jelly and the Rind of a grated Orange or two, stir it up gently over a slow Fire, till all is incorporated together; then take it off, and fill your Clear-cake Glasses, what Scum arises on the Top, you must carefully rake off before they are cold, then put them into the Stove; when you find them begin to crust upon the upper Side, turn them out upon Squares of Glasses, and put them to dry again; when they begin to have a tender Candy, cut them into Quarters, or what Pieces you please, and let them dry till hard, then turn them on Sieves; when thorough dry, put them up into your Boxes.

Note, As they begin to sweat in the Box, you must shift them from Time to Time, and it will be requisite to put no more than one Row in a Box at the Beginning, till they do not sweat.

Lemon-colour Cakes are made with Lemons, as these.

To make Pomegranate Clear-Cakes.

Draw your Jelly as for the Orange Clear-Cakes, then boil into it the Juice of two or three Pomegranate-seeds, and all with the Juice of an Orange and a Lemon, the Rind of each grated in, then [Pg 18] strain it through a Bag, and to every Pound of Jelly put one Pound and a Quarter boiled till it cracks to help the Colour to a fine Red; put in a Spoonful of Cocheneal, prepared as hereafter directed; then fill your Glasses, and order them as your Orange.

To Prepare Cocheneal.

Take one Ounce of Cocheneal, and beat it to a fine Powder, then boil it in three Quarters of a Pint of Water to the Consumption of one Half, then beat Half an Ounce of Roach Allum, and Half an Ounce of Cream of Tartar very fine, and put them to the Cocheneal, boil them all together a little while, and strain it through a fine Bag, which put into a Phial, and keep for Use.

Note, If an Ounce of Loaf-sugar be boiled in with it, it will keep from moulding what you do not immediately use.

To make Pippin-Knots.

Pare your Pippins, and weigh them, then put them into your preserving Pan; to every Pound put four Ounces of Sugar, and as much Water as will scarce [Pg 19] cover them; boil them to a Pulp, and then pulp them through a Sieve; then to every Pound of the Apples you weighed, take one Pound of Sugar clarified, boil it till it almost cracks, then put in the Paste, and mix it well over a slow Fire, then take it off and pour it on flat Pewter-plates or the Bottoms of Dishes, to the Thickness of two Crowns; set them in the Stove for three or four Hours, then cut it into narrow Slips and turn it up into Knots to what Shape or Size you please; put them into the Stove to dry, dusting them a little, turn them and dry them on the other Side, and when thorough dry, put them into your Box.

Note, You may make them red by adding a little Cocheneal, or green by putting in a little of the following Colour.

To prepare a Green Colour.

Take Gumbouge one Quarter of an Ounce, of Indico and Blue the same Quantity; beat them very fine in a Brass Mortar, and mix with it a Spoonful of Water, so will you have a fine Green; a few Drops are sufficient.

[Pg 20]

To make a Compote of Boonchretien Pears.

Pare your Fruit, and cut them into Slices, scald them a little, squeezing some Juice of Lemon on them in the scalding to keep them white; then drain them, and put as much clarified Sugar as will just cover them, give them a Boil, and then squeeze the Juice of an Orange or Lemon, which you best approve of, and serve them, to Table when cold.

Compote of Baked Wardens.

Bake your Wardens in an earthen Pot, with a little Claret, some Spice, Lemon-peal, and Sugar; when you will use them peal off the Skin and dress them in Plates, either Whole or in Halfs; then make a Jelly of Pippins, sharpened well with the Juice of Lemons, and pour it upon them, and when cold, break the Jelly with a Spoon, so will it look very agreeable upon the red Pears.

[Pg 21]

Zest of China-Oranges.

Pare off the outward Rind of the Oranges very thin, and only strew it with fine Powder-Sugar, as much as their own Moisture will take, dry them in a hot Stove.

To Rock Candy-Violets.

Pick the Leaves off the Violets, then boil some of the finest Loaf-sugar till it blows very strong, which pour into your Candying-Pan, being made of Tin, in the Form of a Dripping-Pan, about three Inches deep; then strew the Leaves of the Flowers as thick on the Top as you can; then put it into a hot Stove for eight or ten Days; when you see it is hard candied, break a Hole in one Corner of it, and drain all the Syrup that will run from it, then break it out, and lay it on Heaps on Plates to dry in the Stove.

To candy Violets whole.

Take the double Violets, and pick off the green Stalk, then boil some Sugar till it blows very strong; throw in the Violets, and boil it till it blows again, [Pg 22] then with a Spoon rub the Sugar against the Side of the Pan till white, then stir all till the Sugar leaves them; then sift them and dry them.

Note, Junquils are done the same Way.

To preserve Angelico in Knots.

Take young and thick Stalks of Angelico, cut them into Lengths of about a Quarter of a Yard, then scald them; next put them into cold Water, then strip off the Skins, and cut them into narrow Slips; then lay them on your preserving Pan, then put to them a thin Sugar, that is, to one Part Sugar as clarified, and one Part Water; then set it over the Fire and let it boil, and set it by till next Day, then turn it in the Pan, and give it another Boil; the Day after drain it and boil the Sugar till it is a little smooth, then pour it on your Angelico, and if it be a good Green boil it no more, if not, heat it again; the Day following boil the Sugar till it is very smooth, and pour it upon your Angelico; the next Day boil your Syrup till it rises to the Top of your Pan, then put your Angelico into your Pan, and pour your Syrup upon it, and keep it for Use.

[Pg 23]

To dry it out.

Drain what Quantity you will from the Syrup, and boil as much Sugar as will cover it till it blows, put in your Angelico, and give it a Boil till it blows again; when cold, drain it, and tie it in Knots and put it into a warm Stove to dry, first dusting it a little; when dry on one Side turn it, and dry the other, then pack it up.

To preserve Angelico in Sticks.

Take Angelico, not altogether so young as the other, cut it into short Pieces about half a Quarter of a Yard, or less, scale it a little, then drain it and put it into a thin Sugar as before; boil it a little, the next Day turn it in the Pan the Bottom upwards, and boil it, so finish it as the other for Knots.

Note, When you will candy it, you must drain it from the Syrup, wash it and candy it as the Orange and Lemon.

[Pg 24]

Angelico-Paste.

Take the youngest and most pithy Angelico you can get, boil it very tender, then drain it, and press out all the Water you possibly can, then beat it in a Mortar to as fine a Paste as may be, then rub it through a Sieve; next Day dry it over a Fire, and to every Pound of this Paste take one Pound of fine Sugar in fine Powder; when your Paste is hot, put in the Sugar, stirring it over a gentle Fire till it is well incorporated; when so done, drop it on Plates long or round, as you shall judge proper; dust it a little and put it into the Stove to dry.

To preserve Apricots Green.

Take the Apricots when about to stone, before it becomes too hard for a Pin easily to press through; pare them in Ribs very neatly because every Stroke of the Knife will be seen; then put them into fair Water as you pare them, then boil them till tender enough to slip easily from your Pin, then drain them, and put them into a thin Sugar, that is to say, one Part Sugar clarified, and one Part Water; boil them a little, then set them by till next Day, then [Pg 25] give them another Boil; the Day after drain them and boil your Syrup a little smooth, and put it to them, giving them a Boil; the next Day boil your Syrup a little smooth and put it upon them without boiling your Fruit; then let them remain in the Syrup four or five Days; then boil some more Sugar till it blows, and add it to them; give all a Boil, and let them be till the Day following; then drain them from the Syrup, and lay them out to dry, dusting them with a little fine Sugar before you put them into the Stove.

To put them up in Jelly.

You must keep them in the Syrup so preserved till Codlins are pretty well grown; take Care to visit them sometimes that they do not sour, which if they do, the Syrup will be lost; by reason it will become muddy, and then you will be obliged to make your Jelly with all fresh Sugar, which will be too sweet; but when Codlins are of an indifferent Bigness, draw a Jelly from them as from Pippins, as you are directed in *p.* 8; then drain the Apricots from the Syrup, boil it and strain it through your Strain-bags; then boil some Sugar (proportionable to your Quantity of Apricots you design to put up) till it blows, [Pg 26] then put in the Jelly and boil it a little with the Sugar, then put in the Syrup and the Apricots, and give them all a Boil together, till you find the Syrup will be a Jelly; then remove them from the Fire, and scum them very well, and put them into your Pots or Glasses, observing as they cool if they be regular in the Glasses to sink, and disperse them to a proper Distance, and when thorough cold to cover them up.

To preserve Green Almonds.

Take the Almonds when pretty well grown, and make a Lye with Wood or Charcoal-Ashes, and Water; boil the Lye till it feels very smooth, strain it through a Sieve and let it settle till clear, then pour off the Clear into another Pan, then set it on the Fire in order to blanch off the Down that is on the Almonds, which you must do in this Manner, *viz.* when the Lye is scalding hot throw in two or three Almonds, and try, when they have been in some Time, if they will blanch; if they will, put in the rest, and the Moment you find their Skins will come off, remove them from the Fire, and put them into cold Water, and blanch them one by one rubbing them with Salt, the better to clean them; when you have so done, wash them in several Waters, the [Pg 27] better to clean them, in short, till you see no Soil in the Water; when you have so done, throw them into boiling Water, and let them boil till very tender, till a Pin will very easily pass through them; then drain them, and put them into clarified Sugar without Water, they being green enough, do not require a thin Sugar to bring them to a Colour, but, on the contrary, if too much heated, they will become too dark a Green; the next Day boil the Syrup, and put it on them; the Day after boil it till it becomes very smooth; the Day following give all a Boil together, scum them, and let them rest four or five Days; then, if you will dry them or put them in Jelly, you must follow the Directions as for green Apricots, *p.* 24.

Note, If you will have a Compose of either, it is but serving them to Table when they are first entered, by boiling the Sugar a little more.

To preserve Goosberries green.

Take the long Sort of Goosberries the latter End of *May* or the Beginning of *June*, before the green Colour has left them; set some Water over the Fire, and [Pg 28] when it is ready to boil, throw in the Goosberries, and let them have a Scald, then take them out and carefully remove them into cold Water, and set them over a very slow Fire to green, cover them very close so that none of the Steam can get out; when you have obtained their green Colour, which will perhaps be four or five Hours, then drain them gently into clarified Sugar, and give them a Heat; set them by, and give them another Heat; this you must repeat four or five Times in order to bring them to a very good green Colour: Thus you may serve them to Table by Way of Compose; if you will preserve them to keep either dry or in Jelly, you must follow the Directions as for green Apricots aforementioned, *p.* 24.

To preserve Goosberries white.

Take the large *Dutch* Goosberries when full grown, but before they are quite ripe; pare them into fair Water, and stone them; then put them into boiling Water, and let them boil very tender, then put them into clarified Sugar in an earthen Pan, and put as many in one Pan as will cover the Bottom; then set them by till next Day, and boil the Syrup a little, and pour it on them; the Day after boil it till [Pg 29] smooth, and pour it on them; the third Day give them a gentle Boil round, by setting the Side of the Pan over the Fire, and as it boils, turning it about till they have had a Boil all over, the Day following make a Jelly with Codlins, and finish them as you do the others, in *p.* 28.

To dry Goosberries.

TO every Pound of Goosberries, when stoned, put two Pounds of Sugar, but boil the Sugar till it blows very strong; then strew in the Goosberries, and give them a thorough Boil, till the Sugar comes all over them, let them settle a Quarter of an Hour, then give them another good Boil, then scum them, and set them by till the next Day; then drain them, and lay them out on Sieves to dry, dusting them very much, and put a good brisk Fire into the Stove; when dry on one Side, turn them and dust them on the other; and when quite dry, put them into your Box.

[Pg 30]

To make Goosberry-Paste.

Take the Goosberries when full grown, wash them, and put them into your preserving Pan, with as much Spring-water as will almost cover them, and boil them very quick all to a Pommish; then strew them on a Hair-sieve over an earthen Pot or Pan, and press out all the Juice; then to every Pound of this Paste, take one Pound and two Ounces of Sugar, and boil it till it cracks; then take it from the Fire and put in your Paste, and mix it well over a slow Fire till the Sugar is very well incorporated with the Paste; then scum it and fill your Paste-Pots, then scum them again, and when cold, put them into the Stove, and when crusted on the Top, turn them, and set them in the Stove again, and when a little dry, cut them in long Pieces, and set them to dry quite; and when so crusted that they will bear touching, turn them on Sieves and dry the other Side, then put them into your Box.

Note, You may make them red or green, by putting the Colour when the Sugar and Paste is all mixed, giving it a Warm altogether.

[Pg 31]

Goosberry Clear-Cakes.

Goosberry Clear-Cakes are made after the same Manner as the Paste, with this Difference only, that you strain the Jelly through the Bag before you weigh it for Use.

To dry Cherries.

Stone your Cherries and weigh them, to eight Pounds of Cherries put two Pounds of Sugar, boil it till it blows very strong: put the Cherries to the Sugar, and heat them by Degrees till the Sugar is thoroughly melted, for when the Cherries come in, it will so cool the Sugar that it will seem like Glew, and should you put it on a quick Fire at first, it will endanger the Burning; when you find the Sugar is all melted, then boil them as quick as possible till the Sugar flies all over them, then scum them, and set them by in an earthen Pan; for where the Sugar is so thin, it will be apt to cancker in a Copper or Brass, or stain in a Silver; the next Day drain them, and boil the Sugar till it rises, then put in your Cherries, and [Pg 32] give them a Boil, scum them and set them by till the next Day, then drain them and lay them out on Sieves, and dry them in a very hot Stove.

To preserve Cherries Liquid.

Take the best Morello Cherries when ripe, either stone them or clip their Stalks; and to every Pound take a Pound of Sugar, and boil it till it blows very strong, then put in the Cherries, and by Degrees, bring them to boil as fast as you can, that the Sugar may come all over them, scum them and set them by; the next Day boil some more Sugar to the same Degree, and put some Jelly of Currants, drawn as hereafter directed; For Example, if you boil one Pound of Sugar, take one Pint of Jelly, put in the Cherries and the Syrup to the Sugar; then add the Jelly, and give all a Boil together; scum them, and fill your Glasses or Pots; take Care as they cool, to disperse them equally, or otherwise they will swim all to the Top.

[Pg 33]

To draw Jelly of Currants.

Wash well your Currants, put them into your Pan, and mash them; then put in a little Water and boil them to a Pommish; then strew it on a Sieve, and press out all your Juice, of which you make the Jelly for all the wet Sweet-meats that are red.

Note, Where white Currant-Jelly is prescribed, it is to be drawn after the same Manner; but observe you strain it first.

To make Cherry-Paste.

Take two Pounds of Morello Cherries, stone them and press the Juice out; dry them in a Pan and mash them over the Fire; then weigh them, and take their Weight in Sugar beaten very fine; heat them over the Fire till the Sugar is well mixed, then dress them on Plates or Glasses, dust them when cold, and put them into the Stove to dry.

[Pg 34]

To dry Currants in Bunches.

Stone your Currants and tie them up in little Bunches, and to every Pound of Currants you must boil two Pounds of Sugar, till it blows very strong, then slip in the Currants, and let them boil very fast, till the Sugar flies all over them; let them settle a Quarter of an Hour, then boil them again till the Sugar rises almost to the Top of the Pan, then let them settle, scum them, and set them by till next Day; then you must drain them, and lay them out, taking Care to spread the Sprigs that they may not dry clogged together: then dust them very much, and dry them in a hot Stove.

To preserve Currants in Jelly.

Stone your Currants, and clip off the black Tops, and strip them from the Stalks, and to every Pound boil two Pounds of Sugar till it blows very strong, then slip in the Currants, and give them a quick Boil, then take them from the Fire and let them settle a little; then give them another Boil, and put in a Pint of Currant-Jelly, drawn [Pg 35] as directed in *p.* 33; boil all well together, till you see the Jelly will flake from the Scummer; then remove it from the Fire, and let it settle a little; then scum them, and put them into your Glasses; but as they cool, take Care to disperse them equally.

To preserve Violet-Plumbs.

Violet Plumbs are a long Time Yellow, and are ripe in the Month of *June*, which are preserved as follows; put them into clarified Sugar, just enough to cover them, and boil them pretty quick; the next Day boil them again as before; the Day after drain them again, and take away their Skins, which you will find all flown off, then put them into a Sugar, boil it till it blows a little, give them a Boil; the Day following boil some more Sugar till it blows a little, give them a Boil; the next Day boil some more Sugar to blow very strong, put the Plumbs in the Syrup, boil a little, and scum them; the next Day drain them, and lay them out to dry, but dust them before you put them into the Stove.

[Pg 36]

To preserve Orange-Flowers.

Take the Orange-Flowers just as they begin to open, put them into boiling Water, and let them boil very quick till they are tender, putting in a little Juice of Lemons as they boil, to keep them white; then drain them and dry them carefully between two Napkins; then put them into a clarified Sugar, as much as will cover them; the next Day drain the Syrup, and boil it a little smooth; when almost cold, pour it on the Flowers; the Day after you may drain them and lay them out to dry, dusting them a very little.

To put them in Jelly.

After they are preserved, as before directed, you must clarify a little more Sugar, with Orange-Flower-Water, and make a Jelly of Codlins, which, when ready, put in the Flowers Syrup and all; give them a Boil, scum them, and put them into your Glasses or Pots.

[Pg 37] *Note*, When you boil the Syrup, you must add Sugar if it wants, as well in the Working the foregoing Fruits, as these.

To make Orange-Flower-Cakes.

Take four Ounces of the Leaves of Orange-Flowers, put them into fair Water for about an Hour, then drain them and put them between two Napkins, and with a Rolling-pin roll them till they are bruised; then have ready boiled one Pound of Double-refined-sugar to a bloom Degree; put in the Flowers, and boil it till it comes to the same Degree again, then remove it from the Fire, and let it cool a little; then with a Spoon grind the Sugar to the Bottom or Sides of the Pan, and when it becomes white, pour it into little Papers or Cards, made in the Form of a Dripping-pan; when quite cold, take them out of the Pans, and dry them a little in a Stove.

[Pg 38]

To make Orange-Flower-Paste.

Boil one Pound of the Leaves of Orange-Flowers very tender; then take two Pounds and two Ounces of double-refined Sugar in fine Powder; and when you have bruised the Flowers to a Pulp, stir in the Sugar by Degrees over a slow Fire till all is in and well melted; then make little Drops and dry them.

To preserve Apricots whole.

Take the Apricots when full grown, pare them, and take out their Stones; then have ready a Pan of boiling Water, throw them into it, and scald them till they rise to the Top of the Water; then take them out carefully with your Scummer, and lay them on a Sieve to drain; then lay them in your preserving Pan, and put over them as much Sugar boiled to blow as will cover them, give them a Boil round, by setting the Pan half on the Fire, and turning it about as it boils; then set it full on the Fire, and let it have a covered Boiling; then let them settle a Quar [Pg 39] ter of an Hour, and pick those that look clear to one Side, and those that do not to the other; then boil that Side that is not clear till they become clear; and as they do so, pick them away, lest they boil to a Paste; when you see they look all alike, give them a covered Boiling, scum them, and set them by; the next Day boil a little more Sugar to blow very strong, put it to the Apricots, and give them a very good Boil, then scum them, and cover them with a Paper, and put them into a Stove for two Days; then drain them, and lay them out to dry, first dusting the Plates you lay them on, and then the Apricots, extraordinary well, blowing off what Sugar lies white upon them, then put them into a very warm Stove to dry, and when dry on one Side, turn and dust them again; and when quite dry, pack them up.

Note, In the turning them you must take Care there be no little Bladders in them, for if there be, you must prick them with a Point of a Pen-knife, and squeeze them out, otherwise they will blow and sour.

[Pg 40]

To preserve Apricot-Chips.

Split the Apricots, and take out the Stones, then pare them, and turn them into a circular form with your Knife; then put them into your Pan without scalding, and put as much Sugar boiled very smooth as will cover them, then manage them on the Fire as the whole Apricots, scum them, and set them in the Stove; the next Day boil some more Sugar, to boil very strong, then drain the Syrup from the Apricots, and boil it very smooth; then put it to the fresh Sugar, and give it a Boil; then put in the Apricots and boil them first round, and then let them have a covered Boil, scum them, and cover them with a Paper; then put them into the Stove for two or three Days, drain them, and lay them out to dry, first dusting them.

To preserve Apricots in Jelly.

Pare and stone your Apricots, then scald them a little, and lay them in your Pan, and put as much clarified Sugar to them as will cover them; the next Day drain the Syrup, and boil it smooth, then [Pg 41] slip in your Apricots, and boil as before; the next Day make a Jelly with Codlins, boiling some Apricots amongst them, to give a better Taste; when you have boiled the Jelly to its proper Height, put in the Apricots with their Syrup, and boil all together; when enough, scum them very well, and put them into your Glasses.

To make Apricot-Paste.

Boil some Apricots that are full ripe to a Pulp, and rub the Fine of it thro' a Sieve; and to every Pound of Pulp take one Pound and two Ounces of fine Sugar, beaten to a very fine Powder; heat well your Paste, and then, by Degrees, put in your Sugar; when all is in, give it a thorough Heat over the Fire, but take Care not to let it boil; then take it off and scrape it all to one Side of the Pan, let it cool a little, then with a Spoon lay it out on Plates in what Form you please, then dust them, and put them into the Stove to dry.

[Pg 42]

To make Apricot Clear-Cakes.

First, draw a Jelly from Codlins, then boil in that Jelly some very ripe Apricots, which press upon a Sieve over an earthen Pan, then strain it through your Jelly-bag; and to every Pound of Jelly take the like Quantity of fine Loaf-sugar, which clarify, and boil till it cracks; then put in the Jelly, and mix it well, then give it a Heat on the Fire, scum it and fill your Glasses; in the Drying, order them as has been already directed in *p.* 16.

To make Jam of Apricots.

Pare the Apricots, and take out the Stones, break them, and take out the Kernels, and blanch them; then to every Pound of Apricots boil one pound of Sugar till it blows very strong, then put in the Apricots, and boil them very brisk till they are all broke, then take them off, and bruise them well, put in the Kernels and stir them all together over the Fire, then fill your Pots or Glasses with them.

[Pg 43] *Note*, If you find it too sweet, you may put in a little White-Currant-Jelly to sharpen it to your Liking.

To preserve Rasberries Liquid.

Take the largest and fairest Rasberries you can get, and to every Pound of Rasberries take one Pound and a Half of Sugar, clarify it, and boil it till it blows very strong; then put in the Rasberries, and let them boil as fast as possible, strewing a little fine beaten Sugar on them as they boil; when they have had a good Boil, that the Sugar rises all over them, take them from the Fire, and let them settle a little, then give them another Boil, and put to every Pound of Rasberries half a Pint of Currant-Jelly; let them have a good Boil, till you perceive the Syrup hangs in Fleeks from your Scummer; then remove them from the Fire, take off the Scum, and put them into your Glasses or Pots.

Note, Take Care to remove what Scum there may be on the Top; when cold, make a little Jelly of Currants, and fill up the Glasses; then cover them with Paper first wet in fair Water, and [Pg 44] dry'd a little betwixt two Cloths, which Paper you must put close to the Jelly; then wipe clean your Glasses, and cover the Tops of them with other Paper.

To make Rasberry-Cakes.

Pick all the Grubs and spotted Rasberries away; then bruise the rest, and put them on a Hair-sieve over an earthen Pan, putting on them a Board and Weight to press out all the Water you can; then put the Paste into your preserving Pan, and dry it over the Fire, till you perceive no Moisture left in it, that is, no Juice that will run from it, stirring it all the Time it is on the Fire to keep it from burning; then weigh it, and to every Pound take one Pound and two Ounces of Sugar, beat to a fine Powder, and put in the Sugar by Degrees; when all is in, put it on the Fire, and incorporate them well together; then take them from the Fire and scrape it all to one Side of the Pan; let it cool a very little, then put it into your Moulds; when quite cold, put them into your Stove without dusting it, and dry it as other Sorts of Paste.

[Pg 45] *Note*, You must take particular Care that your Paste doth not boil after your Sugar is in; for if it does, it will grow greasy and never dry well.

To make Rasberry Clear-Cakes.

Take two Quarts of ripe Goosberries, or white Currants, and one Quart of red Rasberries, put them into a Stone-Jug and stop them close; then put it into a Pot of cold Water, as much as will cover the Neck of the Jug; then boil them in that Water till all comes to a Paste, then turn them out in a Hair-sieve, placed over a Pan, press out all the Jelly and strain it thro' the Jelly-bag; to every Pound of Jelly take twenty Ounces of Double-refined Sugar, and boil it till it will crack in the Water; then take it from the Fire and put in your Jelly, stirring it over a slow Fire, till all the Sugar is melted; then give it a good Heat till all is incorporated; then take it from the Fire, scum it well, and fill your Clear-cake-glasses; then take off what Scum is on them, and put them into the Stove to dry, observing the Method directed in *p*. 16.

[Pg 46] *Note*, In filling out your Clear-cakes and Clear-pastes, you must be as expeditious as possible, for if it cools it will be a Jelly before you can get it into them.

White Rasberry Clear-cakes are made after the same Manner, only mixing white Rasberries with the Goosberries in the Infusion.

To make Rasberry Clear-Paste.

Take two Quarts of Goosberries, and two Quarts of red Rasberries, put them in a Pan, with about a Pint and an Half of Water; boil them over a very quick Fire to a Pommish, then throw them upon an earthen Pan, and press out all the Juice; then take that Juice and boil in it another Quart of Rasberries, then throw them on a Sieve, and rub all through the Sieve that you can; then put in the Seeds and weigh the Paste, and to every Pound take twenty Ounces of fine Loaf-sugar, boiled, when clarified, till it cracks, then remove it from the Fire, and put in your Paste, mix it well, and set it over a slow Fire, stirring it till all the Sugar is melted, and [Pg 47] you find it is become a Jelly; then take it from the Fire and fill your Pots or Glasses, whilst very hot, then scum them and put them into the Stove; observe, when cold, the drying them, as in *p.* 16.

To make Rasberry-Biscakes.

Press out the Juice, and dry the Paste a little over the Fire, then rub all the Pulp through a Sieve; then weigh, and to every Pound take eighteen Ounces of Sugar, sifted very fine, and the Whites of four Eggs, put all in the Pan together, and with a Whisp beat till it is very stiff, so that you may lay it in pretty high Drops; and when it is so beaten, drop it in what Form you please on the back Sides of Cards, (Paper being too thin, it will be difficult to get it off;) dust them a little with a very fine Sugar, and put them into a very warm Stove to dry; when they are dry enough, they will come easily from the Cards; but whilst soft, they will not stir; then take and turn then on a Sieve, and let them remain a Day or two in the Stove; then pack them up in your Box, and they will, in a dry Place, keep all the Year without shifting.

[Pg 48]

To make Currant-Paste.

Wash well your Currants and put them into your preserving Pan, bruise them, and with a little Water, boil them to a Pulp, press out the Juice, and to every Pound take twenty Ounces of Loaf-sugar, boil it to crack; then take it from the Fire, and put in the Paste; then heat it over the Fire, take off the Scum, and put it into your Paste-pots or Glasses, then dry and manage them as other Pastes.

To make Rasberry-Jam.

Press out the Water from the Rasberries; then to every Pound of Rasberries take one Pound of Sugar, first dry the Rasberries in a Pan over the Fire, but keep them stirring, lest they burn; put in your Sugar, and incorporate them well together, and fill your Glasses or Pots, covering them with thin white Paper close to the Jam, whilst it is hot; and when cold, tie them over with other Paper.

[Pg 49]

To preserve Peaches whole.

Take the *Newington* Peach, when full ripe, split it, and take out the Stone, then have ready a Pan of boiling Water, drop in the Peaches, and let them have a few Moments scalding; then take them out, and put them into as much Sugar, only clarified, as will cover them, give them a Boil round, then scum them and set them by till the next Day; then boil some more Sugar to blow very strong, which Sugar put to the Peaches, and give them a good Boil, scum them, and set them by till the Day following; then give them another good Boil, scum them and put them into a warm Stove for the Space of two Days; then drain them, and lay them out one half over the other, dust them and put them into the Stove; the next Day turn them and dust them, and when thorough dry, pack them up for Use.

[Pg 50]

To preserve Peach-Chips.

Pare your Peaches, and take out the Stones, then cut them into very thin Slices, not thicker than the Blade of a Knife; then to every Pound of Chips take one Pound and an Half of Sugar, boiled to blow very strong, then throw in the Chips, and give them a good Boil, then let them settle a little, take off the Scum, and let them stand a Quarter of an Hour, then give them another good Boil, and let them settle as before; then take off the Scum, cover them, and set them by; the next Day drain them, and lay them out Bit by Bit, dust them, and dry them in a warm Stove; when dry on one Side, take them from the Plate with a Knife, and turn them on a Sieve; and then again, if they are not pretty dry, which they generally are.

To put them in Jelly.

Draw a Jelly from Codlins, and when they are boiled enough, take as much Jelly as Sugar, boil the Sugar to blow very strong, then put in the Jelly, give it a Boil [Pg 51] and put it to the Chips; give all a Boil and scum them, then put them into your Glasses.

To preserve Walnuts White.

Take the largest *French* Walnuts, when full grown, but before they are hard, pare off the green Shell to the White, and put them into fair Water; then throw them into boiling Water, and boil them till very tender; then drain them and put them into a clarified Sugar, give them a gentle Heat; the next Day boil some more Sugar to blow, and put it to them, giving them a Boil; the next Day boil some more Sugar to blow very strong, put it to the Walnuts, give them a Boil, scum them, and put them by, then drain them and put them on Plates, dust them and put them into a warm Stove to dry.

[Pg 52]

To preserve **Walnuts Black.**

Take of the smaller Sort of Walnuts, when full grown, and not shelled; boil them in Water till very tender, but not to break, so they will become black; then drain them, and stick a Clove in every one, and put them into your preserving Pan, and if you have any Peach Syrup, or of that of the white Walnuts, it will be as well or better than Sugar; put as much Syrup as will cover the Walnuts, boil them very well, then scum them and set them by; the next Day boil the Syrup till it becomes smooth, then put in the Walnuts and give them another good Boil; the Day after drain them and boil the Syrup till it becomes very smooth, adding more Syrup, if Occasion; give all a Boil, scum them, and put them in your Pot for Use.

Note, These Walnuts are never offered as a Sweet-meat, being of no Use but to purge gently the Body, and keep it open.

[Pg 53]

To preserve Nectarines.

Split the Nectarines, and take out the Stones, then put them into a clarified Sugar; boil them round, till they have well taken Sugar; then take off the Scum, cover them with a Paper and set them by; the next Day boil a little more Sugar till it blows very strong, and put it to the Nectarines, and give them a good Boil; take off the Scum, cover them, and put them into the Stove; the next Day drain them and lay them out to dry, first dusting them a little, then put them into the Stove.

To preserve green Amber-Plumbs.

Take the green Amber-Plumbs, when full grown, prick them in two or three Places, and put them into cold Water; then set them over the Fire to scald, in which you must be very careful not to let the Water become too hot, lest you hurt them; when they are very tender, put them into a very thin Sugar, that is to say, one Part Sugar, and two Parts Water; give them a little Warm in this Sugar, and cover [Pg 54] them over; the next Day give them a Warm again; the third Day drain them and boil the Syrup, adding a little more Sugar; then put the Syrup to the Plumbs, and give them a Warm; the next Day do the same; the Day following boil the Syrup till it becomes a little smooth, put in the Plumbs and give them a Boil; the Day after boil the Syrup till very smooth, then put it to the Plumbs, cover them, and put them into the Stove; the next Day boil some more Sugar to blow very strong, put it to the Fruit and give all a Boil, then put them into the Stove for two Days; then drain them and lay them out to dry, first dusting them very well, and manage them in the Drying as other Fruits.

Note, If you find them shrink when first you put them into Sugar, you must let them lie in that thin Syrup three or four Days, till they begin to work; then casting away that Syrup, begin the Work as already set down.

To preserve Green Orange-Plumbs.

Take the green Orange-Plumbs, when full grown, before they turn, prick them with a fine Bodkin, as thick all over as possible you can; put them into cold [Pg 55] Water as you prick them, when all are done, set them over a very slow Fire, and scald them with the utmost Care you can, nothing being so subject to break, for if the Skin flies they are worth nothing; when they are very tender, take them off the Fire and set them by in the same Water for two or three Days; when they become sour, and begin to float on the Top of the Water, be careful to drain them very well; then put them in single Rows in your preserving Pan, and put to them as much thin Sugar as will cover them, that is to say, one Part Sugar, and two Parts Water; then set them over the Fire, and by Degrees warm them till you perceive the Sourness to be gone, and the Plumbs are sunk to the Bottom, set them by; and the next Day throw away that Syrup, and put to them a fresh Sugar, of one Part Sugar, and one Part Water; in this Sugar give them several Heats, but not to boil, lest you burst them; then cover them, and set them in a warm Stove that they may suck in what Sugar they will; the next Day drain the Sugar, and boil it till it becomes smooth, adding some more fresh Sugar; pour this Sugar on them, and return them into the Stove; the next Day boil the Syrup to become very smooth, and pour it upon [Pg 56] your Plumbs, and give all a gentle Boil, scum it and put them into the Stove; the Day following drain them out of that Syrup, and boil some fresh Sugar, as much as you judge will cover them, till very smooth put it to your Plumbs, and give all a very good covered Boiling; then take off the Scum and cover them, let them stand in the Stove two Days, then drain them and lay them out to dry, dusting them very well.

To preserve the green Mogul-Plumb.

Take this Plumb when just upon the turning ripe, prick with a Pen-knife to the very Stone on that Side where the Cleft is, put them into cold Water as you do them, then set them over a very slow Fire to scald; when they are become very tender, take them carefully out of the Water and put them into a thin Sugar, that is, half Sugar, and half Water, warm them gently, then cover them, and set them by; the next Day give them another Warm and set them by; the Day following drain their Syrup and boil it smooth, adding to it a little fresh Sugar, and give them a gentle Boil, the Day after boil the Sugar very [Pg 57] smooth, pour it upon them and set them in the Stove for two Days; then drain them and boil a fresh Sugar to be very smooth, or just to blow a little, put it to your Plumbs and give them a good covered Boiling; then scum them and put them into the Stove for two Days, then drain them and lay them out to dry, dusting them very well.

To preserve the Green Admirable-Plumb.

This is a little round Plumb, about the Size of a Damson; it leaves the Stone, when ripe, is somewhat inclining to a Yellow in Colour, and very well deserves its Name, being of the finest Green when done, and with the tenth Part of the Trouble and Charge, as you will find by the Receipt.

Take this Plumb, when full grown, and just upon the Turn, prick them with a Pen-knife in two or three Places, and scald them, by Degrees, till the Water becomes very hot, for they will even bear boiling; continue them in the Water till they become green, then drain them, and put them into a clarified Sugar, boil them [Pg 58] very well, then let them settle a little, and give them another Boil; if you perceive they shrink and take not the Sugar in very well, prick them with a Fork all over as they lie in the Pan, and give them another Boil, scum them, and set them by; the next Day boil some other Sugar till it blows, and put it to them, and give them a good Boil, then scum them and set them in the Stove for one Night; the next Day drain them and lay them out, first dusting them.

To preserve yellow Amber-Plumbs.

Take these Plumbs, when full ripe, put them into your preserving Pan, and put to them as much Sugar as will cover them, and give them a very good Boil; then let them settle a little, and give them another Boil three or four Times round the Fire, scum them, and the next Day drain them from the Syrup, and return them again into the Pan, and boil as much fresh Sugar as will cover them to blow; give them a thorough Boiling, and scum them, and set them in the Stove twenty-four Hours; then drain them, and lay them out to dry, after having dusted them very well.

[Pg 59] *Note*, In the scalding of green Plumbs, you must always have a Sieve in the Bottom of your Pan to put your Plumbs in, that they may not touch the Bottom, for those that do, will burst before the others are any thing warm.

To put Plumbs in Jelly.

Any of these Sorts of Plumbs are very agreeable in Jelly, and the same Method will do for all as for one: I might make some Difference which would only help to confound the Practitioner, and thereby swell this Treatise in many Places; but, as I have promised, so I will endeavour to lay down the easiest Method I can to avoid Prolixity, and proceed as above, *viz.*

[Plumbs in Jelly.] When your Plumbs are preserved in their first Sugar, and you have drained them in order to put them in a second, they are then fit to be put up Liquid, which you must do thus: Drain the Plumbs, and strain the Syrup through a Bag; then make a Jelly of some ripe Plumbs and Codlins together, by boiling them in just as much Water as will cover them, press out the Juice and strain it, and to every Pint of [Pg 60] Juice boil one Pound of Sugar to blow very strong, put in the Juice and boil it a little; then put in the Syrup and the Plumbs, and give all a good Boil; then let them settle a little, scum them and fill your Glasses or Pots.

To preserve green Figs.

Take the small green Figs, slit them on the Top, and put them in Salt and Water for ten Days, and make your Pickle as follows.

Put in as much Salt into the Water as will make it bear an Egg, then let it settle, take the Scum off, and put the clear Brine to the Figs, and keep them in Water for ten Days; then put them into fresh Water, and boil them till a Pin will easily pass into them; then drain them and put them into other fresh Water, shifting them every Day for four Days; then drain them, and put them into a clarified Sugar; give them a little Warm, and let them stand till the next Day; then warm them again, and when they are become green give them a good Boil, then boil some other Sugar to blow, put it to them, and give them another good Boil; the next Day drain them and dry them.

[Pg 61]

To preserve ripe Figs.

Take the white Figs, when ripe, slit them in the Top, and put them into a clarified Sugar, and give them a good Boil; then scum them, and set them by; the next Day boil some more Sugar till it blows, and pour it upon them, and boil them again very well, scum them and set them in the Stove; the Day after drain them and lay them out to dry, first dusting them very well.

To preserve green Oranges.

Take the green Oranges and slit them on one Side, and put them into a Brine of Salt and Water, as strong as will bear an Egg, in which you must soak them at least fifteen Days; then drain them and put them into fresh Water, and boil them tender; then put them into fresh Water, again, shifting them every Day for five Days together; then give them another Scald, and put them into a clarified Sugar; then give them a Boil, and set them by till next Day, then boil them again; the next [Pg 62] Day add some more Sugar, and give them another Boil; the Day after boil the Syrup very smooth and pour it on them, and keep them for Use.

Note, That if at any Time you perceive the Syrup begin to work, you must drain them, and boil the Syrup very smooth and pour it on them; but if the first prove sour, you must boil it likewise. Green Lemons are done after the same Manner.

Note also, If the Oranges are any thing large, you must take out the Meat from the inside.

To preserve green Grapes.

Take the largest and best Grapes before they are thorough ripe, stone them and scald them, but let them lie two Days in the Water they were scalded in; then drain them and put them into a thin Syrup, and give them a Heat over a slow Fire; the next Day turn the Grapes in the Pan and warm them again; the Day after drain them and put them into a clarified Sugar, give them a good Boil, [Pg 63] and scum them, and set them by; the following Day boil some more Sugar to blow, and put it to the Grapes, and give them a good Boil, scum them and set them in a warm Stove all Night; the next Day drain them and lay them out to dry, first dusting them very well.

To preserve Bell-Grapes in Jelly.

Take the long, large Bell, or Rouson-Grapes, and pick them from the Stalks, then Stone them and put them in boiling Water, and give them a thorough Scald; then take them from the Fire and cover them close down, so that no Steam can come out; then set them upon a very gentle Fire, so as not to boil for two or three Hours; then take them out, and put them into a clarified Sugar boiled, till it blows very strong, as much Sugar as will a little more than cover them; then give them a good Boil and let them settle a little: then give them another Boil, scum them, and then boil some other Sugar to blow very strong; and take as much Plumb-Jelly as Sugar, and give all a Boil, then add to it the Grapes, and give them all a Boil together, scum them well, and put them up into your Pots or Glasses.

www.ingramcontent.com/pod-product-compliance
Lightning Source LLC
Chambersburg PA
CBHW031631210526
45464CB00004B/1853